Bee

(Backyard Beekeeping)

Essential Beginners Guide to Build and Care

For Your First Bee Colony and Make Delicious Natural Honey From Your Own Garden

Table of Contents

Introduction 1

Chapter 1: History of Beekeeping 5

Chapter 2: Starting Your Apiary 10

    Building Supers and Frames 14

    Assembling The Hive 20

    Preparing Food Supplements and Medications 25

Chapter 3: Bee Biology, Equipment, and Bee Installation 29

    Equipment for Beekeeping 32

    Receiving and Installing Packaged Bees 35

    Releasing Queens 38

    Bee Stings 40

    Medications 42

Chapter 4: Bee Management 43

    Hive Maturation 44

    The Brood Nest 46

Chapter 5: Harvest Time 53

    Harvesting and Processing 54

Extracting Honey 55

Conclusion 58

Introduction

A beekeeper, also known as an apiarist, creates a system for honey bee collection. But contrary to popular belief, it's not just honey that they get to harvest. Beekeepers can also harvest pollen, propolis, royal jelly, and beeswax. Beekeepers use hive products to aid in the pollination of crops. Aside from that, hive products are also known for their endless health benefits. Beekeepers can also breed bees to be sold to other beekeepers. A bee yard, or an apiary, is where bees are kept.

There are 140,000 beekeepers in the United States keeping 3.2 million beehives. American beekeepers produced 200 million pounds of honey a year. In addition, honey bees pollinate and make possible many of the fruits and vegetables that make up the American diet. In fact, the annual contribution of honey bees to crop pollination and production is over nine billion dollars.

Beekeepers earn substantial amount of money for keeping bees. It is a business that can be easily sustained by following the universal rules for satisfactory bee production. These rules will be discussed in detail in this book. Aside from honey production, keeping bees is also a good decision if you have a garden. Aside from harvesting honey and other hive products, your garden will also thrive with the aid of bees for pollination.

But to tell the truth, those are not the reasons why many beekeepers have kept bees. Most of them keep bees because they enjoy the outdoors. They enjoy helping a young colony grow in spring. They enjoy watching the bees make honey. They enjoy harvesting, bottling, and selling honey as well. They enjoy contributing to biodiversity and encouraging pollination.

Bee keeping is considered more than just a business; it is a hobby that you and your family can enjoy. Bee keeping can be a highly profitable and unique pastime when done properly.

If you raise animals or work with wood, grow a garden or have an entrepreneurial spirit, you might be a candidate for beekeeping. Raising a huge colony of bees can be pretty satisfying. Bees can double in number by the next spring when they are raised properly. An expert beekeeper can raise enough bees to be sold to other beekeepers.

Beekeeping is also a great complementary hobby if you work well with wood. If you are a wood hobbyist, then you can create your own honey super – the collection mechanism for your beehive. You can customize it based on your preferences.

As a beginner, you would want to be as hands on as possible so you can be in tune with your honey bee production. It is also a very rewarding experience when you get to create your own frame. As mentioned above, beekeeping is very beneficial if you are growing a garden. The United States Department of Agriculture has pointed out how bee pollination is able to increase crop value by an additional of $15 billion dollars. This is particularly true for specialty crops like fruits, vegetables, berries, and nuts. The direct and indirect benefits of beekeeping have proven to be phenomenal.

Bees are very diligent workers. They are independent workers and require very little supervision. Once you have setup the initial requirements for beekeeping, you will only need at least half an hour every week for planning. Twice a year, you will need to spend more time with them for honey collecting. Other than that, you do not need to devote so much of your time to beekeeping.

But there's one thing that seem to keep people away from bee keeping – the idea of getting stung. Bee sting is one of the main reasons why a lot of people are reluctant to start a beekeeping business, but the truth is, bees are gentle creatures, at least until they sense danger to their hive. You'd be surprised at how gentle they can be, and of the many places where you can keep bees. You do not need a very large place to store your honey supers. What is even more surprising, you can even keep your honey supers in your backyard. In this book, we'll go ahead and discuss the basics of backyard beekeeping. As an example, we'll set up ten bee hives from scratch--that is with packages of bees in the mail, just like what veteran beekeepers did when they were starting out.

We'll follow the progress of this bee yard, or apiary, as it matures and develops in to a full blown colony, so you will see what you can expect. This book can be your guide to starting your apiary. But before we begin, let's look at some other history of beekeeping so you can have a better appreciation as to how the science of beekeeping started.

Chapter 1: History of Beekeeping

For centuries, honey and fruit were mankind's only sweets. Therefore, bees and honey figured prominently in early societies. The Christian Bible is full of references to bees and honey. Beeswax was used in art, and a weatherproof cloth, rustproof metal, and make writing tablets.

There are several opinions as to when bees came to existence. According to some sources, bees were first known to exist 40 million years ago. However, there are claims that honey bee fossils were found and the fossilized remains dated back to as far as 150 million years ago.

According to the earliest records, humans started to consume honey 10,000 years ago. This was evidenced by the historical drawings in caves. In Spain, drawings of beekeeping practice date back to between 6,000 and 8,000 years ago. In Eastern Spain, you can find the Cuevas de la Araña (or the Spider Caves) whose rock art depicts an image of a person gathering honey from a tree. These groups of caves are discovered in Bicorp, Valencia by Jaime Gari i Poch, a local teacher. These rock art was among the Rock Art of the Iberian Mediterranean Basin of the World Heritage Site.

At first, people simply hunted bees in hollow trees or cave, and destroyed their nest to get the honey and beeswax. But by 2400 B.C., ancient Egyptians had learned to keep bees in clay pots, and probably did not kill their colonies at harvest time. Ancient Greeks and Romans kept bees. It was also at this time that fire and smoke were discovered. Beekeepers during the Niuserra dynasty were illustrated to be blowing smoke into the beehives while the honeycombs were removed. The honey they have collected were stored in earthen jars. Their philosophers wrote romantic, but biologically inaccurate, accounts of bees and their ways.

Rome maintained commercial apiaries in Spain. During the dark ages, there was little advancement in beekeeping. However, in monasteries--the only oasis of literacy at that time--bees were kept, honey was produced, beeswax candles were made, and the ancient bee text of the Greeks and Romans were preserved.

During the middle ages, beginning about 1500 years ago, beekeepers started cutting trees that contain bee nests and keeping the long sections as hives. At this time, the well-known straw skep appeared. During Europe's discovery of the Americas, honey bees followed shortly thereafter. We don't know when or where bees were introduced to North America, but by the 1640s they were well established along the eastern coast.

In 1682, an English clergyman, named George Wheler, documented the Greek hives he saw in one of his travels. The Greek hives became the basis for the creation of the hives today that features movable frames.

In 1700, Claire Preston explained in her book "Bee" how bees gather the nectar from the flowers and how the nectar was made into honey. Before this, it was believed that bees directly collected the honey from the flowers without involving other processes.

A Polish apiculturist, named Johann Dzierzon, devised the first-ever movable comb hive in 1838. This allowed the harvesting of the honey in each individual combs while preserving the composition of the hive. Dzierzon also discovered the parthenogenesis phenomenon in bees that allows them to reproduce asexually.

Many of the most profound discoveries in beekeeping were made in the 19th century. In the early part of that century, there was a rash of experimentation to invent the perfect hive, resulting in some outlandish designs. But then, the American Lorenzo Lorraine Langstroth observed that bees in nature maintain a space of one quarter-inch to three-eighths of an inch around their combs. Within this space, bees will not build combs. Instead, they leave it open to allow movement in the hive. By using this principle of bee space, Langstroth built the first hive with movable and interchangeable combs. This breakthrough made possible the rapid growth and standardization of beekeeping in the United States. Langstroth's invention was quickly followed by wax foundation for building standard sized combs, a centrifugal honey extractor, and a bellow smoker to calm the bees. Langstroth's observation was built around the works of Dzierzon. Langstroth was credited for the discovery of the "bee space" although this has been implemented in the earlier works of Dzierzon. Langstroth industrialized the beekeeping business that is why he is named the "Father of American Beekeeping

In 1890, English beekeeper and inventor named William Broughton Carr created the WBC hive. The WBC hive resembles the classic hive as it has a double-walled insulation for the bees. It has an external housing that covers the hive inside.

In 1948, Abbé Warré published the book "Beekeeping for All" that features a top bar hives. This hive is a frameless, single-story hive that has combs hanging on removable bars. He also advocated the practice of having far less interference with bees and their hives.

After over 100 years, these same technologies--but only small changes--are the industry standards today, which is a testament to the enduring genius of their inventors. So let's go ahead and start establishing our apiary. Most of the challenges encountered by a beginning beekeeper occur during the first year in the practice. There's a lot to be done during your first year of beekeeping.

Chapter 2: Starting Your Apiary

Building your own apiary offers the sweetest rewards especially if you have a garden. Keeping an apiary is just like having a garden; the time and effort you spend with your bees is just the same as with your produce. If you have a garden, then keeping an apiary should be as easy for you.

Along with your produce, you can harvest honey and other hive products. Your bees can also help your garden thrive. Bees are among the best pollinators your garden can have. And do not forget the joy of having a new hobby plus the knowledge that you are helping the ecosystem with your new-found pastime.

In starting your apiary, there are several things you need to consider. During the planning stage, the first question you need to answer is where you will keep your bees. There is really no need to think that far. If you have a backyard, then that may be a good place to start. However, you might need to check with your local community first before diving in. You will need to ensure that keeping bees in your backyard is legal. In some counties, there are ordinances in place that regulate beekeeping.

Before installing your hives, you also need to set up plans about installing fences on your backyard first. Make your fence high, at least 8 feet tall will do. Your fence will guide your bees as to how high they should fly so they will not bump into your neighbors and sting them.

Most of the work is done at the planning and building stage. Once you are past these stages, then your job will be easier and your apiary will no longer need much of your time. So now it is time for you to build.

You will need to build or buy a hive for your bees. You need to have your hives ready before your bees arrive. A hive is made up of stacked boxes called *Supers*. A standard hive is composed of three to four *supers*.

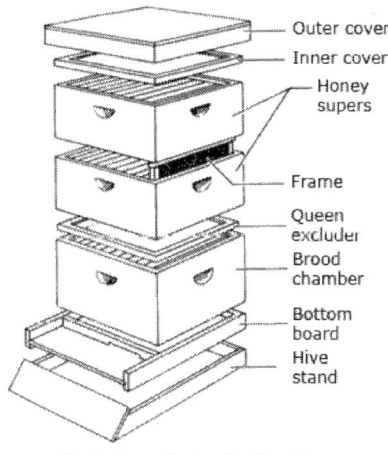

Illustration of a standard beehive

Inside each *super* are ten removable frames that hang vertically. Bees build their combs on these frames. Almost all equipment nowadays is made to standard Langstroth dimensions. Let's talk about the placement of your hives.

You want the entrances of your beehives to face South to Southeast because this provides maximum sun exposure. So, keep this in mind when you are scouting out locations for your bees. Keep your beehives discreet, or you can hide them from sight if needed. Many people are afraid of beehives and may react with alarm or hostility when they see them. This is especially true if you are planning to keep your beehives on your backyard.

Although properly managed beehives are not a public threat, there is still no reason to advertise their presence. Nevertheless, you should always inform your immediate neighbors that you would be keeping beehives in your property; and remember that free honey goes a long way to keep neighbor relations sweet. If you have close neighbors, place your hives so that they are next to a tall barrier--like a fence or a hedge--that forces them to fly above pedestrian level. As mentioned above, a fence that is at least 8 feet in height can be your guide. It is also up to you to make your neighbors feel safe by following safety precautions when it comes to backyard beekeeping.

If water is scarce, provide a source so that bees won't look for water at your neighbor's swimming pool or birdbaths. Your neighbors may not appreciate it if their backyard gets swarmed by your bees so you will need to provide their needs so they will not try to find it elsewhere. When providing a water source, make sure to make use of rocks to keep bees from drowning. Check your local ordinances about beekeeping regulations and register your hives with your state Department of Agriculture. Your county extension agent can help you with these and all of your questions about beekeeping.

Building Supers and Frames

Woodenware comes pre-cut from the manufacturer. *Supers* come in two common sizes: a nine and a half inch deep *super* and a shallower six and five-eighths inch *super*. Once the super has dried, you can go ahead and put it together. The long parts are the sides of the *super*. The shorter parts are the ends.

In assembling your super, you will need a razor blade, a carpenter's square, a hammer, and four clamps. To get a better picture of how your super should look like, lay out all the four sides of your hive. In the ends are dado cuts. These cuts receive the sides. Check for splinters, extra wood, or sharp edges along the finger sides. You can use your razor blade to trim off any excess wood. You need to make sure that all the finger woods are clean and precise so it will be easier to put them together.

The grove that is perpendicular to the dado cuts is called a frame rest. Each side of your hive box should nail holes that are pre-drilled in them. Each finger sides should have a nail hole. Each side has 5 fingers so there should be 5 nail holes. And each super has two sides so there should be 10 holes in all. However, if your hive box did not come with the pre-drilled holes, then it is advisable to drill the holes first before gluing the sides together. Although some may not find this necessary, drilling the holes first will help you avoid having misaligned nails when you hammer the box together. After all these preparations, you can start putting your box together.

First, we use wood glue to attach the sides and ends of the *super*. We use the glue in combination with six penny nails. It is really important before you start nailing that you make sure that your handles are both on the outside.

Hold your box in place by applying one clamp in each side of your hive box until you form the shape of your box. Your clamps should secure the shape of your hive box while you nail the sides. Others may find using clamps unnecessary, the use of clamps will help you nail your box together by keeping all the sides in place. If you do not use clamps, the sides that are not yet secured with nails may fall apart. Using your carpenter's square, you can check if your box is a perfect square. If not, adjust your clamps until they are aligned.

Also, before you nail it down tight, make sure your ends or edges are flushed together. Here's another important step: Make sure that the edge across the top for the frame rest is directly opposite to the edge on the other end. You can now start putting your box together. Choose a top corner and hammer in a nail. On the adjacent side of the same corner where you hammered your first nail, hammer your second nail. Every once in a while, check to see if your box is still aligned. It is not necessary to check after you hammer each nail. However, not doing so may cause you more trouble. It is harder to remove the nails and to start over than to check your box and adjust the clamps when necessary. After you finish hammering two nails on one side, move on to the second corner and just repeat what you have done in the first corner. Then finish hammering the other two corners. You should have hammered in 8 nails when you are finished with all the sides.

After you are done with the top corners, remove your clamps and flip your box over. And then secure the box again with the 4 clamps just like what you did with the top part of your super. Check again if your box is aligned and adjust the clamps and the box if necessary. Hammer one nail in one corner and put in another nail in the adjacent corner. Do these steps for all the remaining three corners. You should have 16 nails in all after you finished all the corners.

Now, you can fasten all the fingers in place by hammering in the nails and working your way on one side until you finish all the sides. You should have 40 nails hammered in after you finish all the sides.

To create a strong corner, put at least four penny nails on each side.

When you finish your deep *super*, the dado cut will be on top. It forms a shelf for the frames to hand on. One of the most common mistakes in building *supers* is to nail the sides in upside down, so that the handles are useless. Coat all exterior surfaces with an oil-based paint. You don't have to use the color white like most *supers*. Any color will do.

However, it is recommended to use blue, green, tan, and brown in addition to white. These colors combined make the hives less noticeable. Compared to the *supers*, the frames that go inside them are a little bit more tedious to build.

Depending on the configuration of your hive and the goals that you are trying to achieve, there are several different types of frame that you can choose from. However, following the Langstroth Model, we will assemble a Groove Bottom Bar or the Wedge Top Bar wooden frame. In a Groove Bottom Bar, the foundation sits inside the groove that runs along the bottom of the frame. As for the Wedge Top Bar, it has a wedge cleat which is a piece of wood that runs along the top bar. Prior to assembly, the wedge cleat should be removed to be reinstalled again after the foundation is set in place.

In assembling the frame, you will need a razor blade, a carpenter's square, a hammer, wood glue, paper towels or rags to clean up glue spills, 1 ¼" frame nails (or the 18 x 1 ¼ wire nails), and the 4 pieces of wood that make up your frame. Before assembling your frame, you need to cut off excess pieces of wood from the frame pieces using your razor blade. If you will be assembling several pieces of frames, it will save you more time if you trim them all first before constructing your frames. When the frame pieces are all clean, you can start assembling your frame.

First, we break off the wedge that's attached to the frame top bar. Using your razor blade, run the blade along the groove on the top bar to remove the wedge cleat. This wedge is later used to fasten the foundation in place. Shave off the little bit of wood that remains after breaking the wedge off. Just like with the *supers*, use wood glue along with nails to make the join strong. Put a small amount of glue in the upper groove on one side of the bar. From where you applied the glue, fit the top of the frame to the side bar. Push them together tightly so the pieces fit together very well. The pieces should fit snugly. Glue both end bars at the same time and fasten them onto the top bar. You should now have a three-sided frame that is made of a top bar and two side pieces. Put the bottom bar of the frame by gluing it to the two side bars. Push it down until the pieces fit well. Using a rag or paper towels, remove any excess glue that may have spilled between the joint pieces. The frame's bottom bar has a split down the middle to accommodate the foundation. Line up all the pieces so they're flush at the edges before nailing the bottom bar in place. It is now time for you to nail your frame together. You do not have to wait for the glue to dry before nailing your frame. You need to nail your frame right away after you glued them together.

When driving your nails into your frame, check for the squareness of your frame using a carpenter's square. If your framed is not square, gentle put the frame in place.

Start hammering your first two nails at the bottom of your frame. The nails should be near where the side bars were attached and they should be parallel to one another. Do the same on the other side of the bottom of the frame. There should now be four nails at the bottom of your frame.

Turn your frame over so you will have the top of the frame in front of you. Drive two nails in the same way that you drove the nails on the bottom of your frame. Hammer two nails again at the other side of your top of your frame.

Then you can now move on the side bars. Hammer two nails at the side adjacent to the top of your frame. Do the same on the other side of the frame.

Now that your frame is put together, don't forget to nail the top bar to the end bars; one nail in the middle will do. Here's another important step that you need to remember: Put another nail from the end bar into the top bar. These nails on the side will provide important support when the frame gets heavy with honey later in the season.

Next, put brass eyelets into the holes in the end bar. The eyelets keep the wire from cutting into the soft wood. Now, the frame is ready for cross wiring. Put a small nail on the end bar. Thread the wire through the holes. There are four holes to accommodate four cross wires. However, an x-shaped configuration is preferred because it provides similar support and uses less wire.

Fasten one end of wire to the nail. Pull the wire tight on the spool end. Once it is tight, fasten the wire permanently by driving down the nail where it is fastened. Cut off any excess wire protruding out from the nail. Next, we have the wax foundation. It is a sheet of beeswax imprinted with the shapes of cells. When it's installed in frames, bees use it literally as foundation on which to build their combs.

Slip a sheet of wax foundation into the frame. Return the wooden wedge that we broke off earlier, making sure it is on top of the wire hooks. After making sure it is flushed at the edges, nail the wedge into place. Finally, embed the cross wires into the wax. Shallow frames are made the same way. The only difference being that the crossed wires are installed into horizontal rows.

Assembling The Hive

Now that we've finished building our frames and *supers*, we can assemble our hive. A typical hive rests on top of two concrete blocks to prevent soil contact with the wood. Always prop the bottom board forward to allow moisture drainage out the entrance. On top of the bottom board go two hive bodies that is the deep *supers*.

These provide egg-length space for the Queen, and they are the heart of the colony. Bees naturally place brood--that is their developing young bees--downward in the brood nest. And then surround them above and into the sides with an insulating layer of honey.

There are two types of covers on beehives. The first is an inner cover, which provides insulation with a dead air space. The second is a heavier outer cover, which is simply a weather barrier. Beekeepers should never harvest honey from the brood nest. Instead we provide space for surplus honey by adding shallow *supers* at the top of the deep *supers*.

As the season progresses, we add more shallow *supers* to accommodate the incoming nectar. These shallow *supers* represent the honey that can be harvested by the beekeeper. For starting a brand new apiary, we only need one hive body per colony. Below is a step by step procedure to assembling your hive.

1. Assemble the board at the bottom of your hive.

The plywood floor should be positioned into the dado groove of the hive's short rail. With the dado side up, the rail should lay on your worktable. Either end of the plywood floor could serve as the rear of your bottom board. On both sides of your plywood floor, put the long rails while fitting the plywood into the dados. The dado should face in the same way so your bottom board will not turn out to be lopsided. Check if the alignment of the rails and the fit match the floor.

After you are certain that they are aligned and that they fit, put a #6 1-3/8" galvanized screw decks halfway towards the center of the three rails. Using a drill, the screws should be driven down the rails into the edges of the plywood. However, do not screw them all the way down right away. Check first if all the parts fit perfectly. Once all the screws are drilled down, there will be no going back. Once you are satisfied with the assembly of all the parts, drill four supplementary evenly spaced screws along the short rail. It will be easier to put in the screws if you drill a 7/64" hole in each of the place where you will fasten your screws. Drilling a hole first will also make you ensure that the wood will not split when you put in your screws. Place the entrance reducer at the entrance of the hive to regulate ventilation and prevent other bees from other colonies from stealing the honey from that hive. Your entrance reducer should remain loose as it is not used all year round.

2. Prepare the supers that you have assembled.

Since you have assembled your supers, you just need to prepare them so you can have them handy as you assemble your hive.

3. Assemble the inner cover.

The plywood cover should be positioned and inserted into the dado of the long rails and the short rails. Imagine putting together a picture frame. All the rails should have the entire groove facing the same direction. If not, your inner cover will be lopsided. Check for the fit and the alignment of your plywood cover and the rails. Drill a deck screw halfway into each corner. Make sure that everything fits properly and it is square before drilling all your screws all the way in. There is no need to paint the inner cover as with all the inner parts of the beehive.

4. Assemble the outer cover.

The plywood should be inserted in the rabbeted groove in one of the lone rail. Do the same thing on the other side of the long rail. Put both the short rails into the plywood board.

The rails should form the frame around the plywood board. Make a stopper that is attached on top of your worktable so you can push your outer cover against while assembling the part. Put the outer cover on your table. Drill two screws into the corners of the short rails. Do the same thing on all corners of the short rails. The stopper can assist you so your cover will not slide on your table as you put in the screws. Fasten the plywood insert into the assembly using deck screws. Force the screws into the rails and into the plywood board. The trick is to fasten them using five screws that are spaced evenly in each long rail and four screws in each short rail. The aluminum flashing should be centered at the top of the outer cover. The flashing should be bent along the edges of the frame. There should be a 7/8" lip surrounding the top of the edge. Repeat this on all the four sides. Bend the corners of the flashing. Just imagine making your bed. Using a rubber mallet, flatten out the corners. Be cautious when handling the aluminum flashing as they are very sharp. You can use work gloves to prevent from being cut. Put the folded edges in place using the #8 x ½" lath screws.

6. Start building your hive using all the assembled parts.

The bottom board should be placed on a flat ground. The bottom board will serve as the floor of the beehive. This is where the entrance is located. You can also use a hive stand to elevate your beehive off the ground. Elevating your hive will have a benefit for both you and the bees. You will not need to bend over when checking on your hive and the bees will have better ventilation.

At the top of the bottom board are the two deep hives. The deep hives serves as the raising hive for the baby bees and for food storage. The lower deep hive is where the baby bees are raised and the upper hive is where they store their food.

The medium super should be placed on top of the deep hives. The medium super is the place for stocking the extra honey, which you can later harvest. The medium super can hold up to 35 pounds of honey.

Some hives have a queen excluder between the medium super and the top deep. The queen excluder acts as the barrier between the queen bee and the medium super where the honey is located. This is to prevent the queen bee from laying her eggs in the medium super. If a queen bee lays eggs in the medium super, other bees will bring in pollen which will affect the purity of the honey.

Depending on how much honey you are getting from your hives, you can add two or more medium supers on your hive.

On top of the medium super is the inner cover that controls the ventilation on the hive. Putting the deeper ledge face up will allow more ventilation into your hive. Putting it face down will reduce the ventilation.

The outer cover serves as the roof of the hive. The outer cover provides protection for your hive. Just put the outer cover into the inner cover and you are set to go. Your hive is ready for your bees.

Preparing Food Supplements and Medications

There are constant discussions as to whether you should give your bees food supplements and medication or not. Some beekeepers like to have a treatment-free apiary. However, this poses a great threat to the bees. If you are truly a beekeeper, you would not want to lose your bees to some pests and diseases. Providing food supplements and medication can help your bees multiply and produce more. To become a good beekeeper, you need to know how your bees can be kept alive and well. You can help your bees get past the challenging times.

In some counties, hives needs to be registered and inspected every 18 months. For the producers of queen and packaged bees, it should be every 12 months. You may need to contact your county extension agent so they can help you with the registration and inspection of your hives.

Before we install the bees after setting up our equipment, we have to mix up some feeds and medication. To encourage fast population growth, we'll feed the colonies with protein supplements in the form of pollen substitute patties. To begin with, we need a mixture of Expeller-processed soy flour and Brewer's yeast.

Next, we'll have to add syrup made of two parts sugar and one part water. Make sure to mix up a lot of syrup since we're going to use the same concentration later to feed the baby bees in liquid form. Mixing the pollen substitute patties is a little touch-and-go. What we're trying to get is a thick, pasty consistency.

Because this stuff is so messy, you should shape the patties using wax paper. We just put it in the paper and press it out in a flat, patty form. You can then freeze the patties and just take them out when you need them.

We're going to have to protect our colony against brood diseases, such as American foul brood (AFB). AFB is a disease that attacks the bee pupae and larvae. The bacteria are spread through the adult bees and equipment that had been contaminated with the disease. The infected larvae become dark brown as opposed to their pearly white color. After the larvae had been capped, they will die. The cappings will sink into the cell floors. You can check if your hive is infected by AFB by shoving a stick into the dead larvae and mix it until you form a mass. Lift the stick and if the mass resembles a string of about 1 inch, then most probably they were killed by AFB.

Another variation of this disease is the European foul brood (EFB). With EFB, the infected larvae die before they reach the capping stage. The dead larvae melt and form at the bottom of the cell floors. EFB is not as dangerous as the AFB and colonies infected with this disease may have the tendency to recover on their own. You can prevent AFB and EFB by applying a treatment to your beehive. To do this, we're going to use the antibiotic Terramycin. To prepare this, we're going to mix a whole 6-ounce package of Terramycin with two and a half pounds of powdered sugar. This will be fed to the bees in a dry state.

We need to ensure that the colonies will not get nosema disease. Nosema is a parasite in the digestive tract which can severely slow colony growth. This is a protozoan disease that affects adult bees. Colonies infected by this disease may not have the ability to build up their colony. Bees of infected colonies become weak and they crawl at the front of the hive. You can prevent Nosema by choosing a hive that has a good flow of air. Nosema is more prevalent in damp and cold states. You can treat Nosema with Fumidil-B. Fumidil-B is an antibiotic which can protect the colonies from nosema disease. Following the label instructions for ten colonies, we'll mix 1 gram of Fumidil-B into one quart warm water.

Once it is mixed, pour the medication mixture into the tank of sugar syrup and then stir. When mixing this medication, make sure that the total volume is equal to the number of colonies that you have. This will ensure that each colony have at least a gallon of syrup each. The syrup medication mixture goes into court jars. The perforated lids of the jars fit into the Boardman feeders at the entrance of the hives.

Our young colonies will get this sugar syrup and Fumidil-B mixture as soon as they're installed into the new hives. The syrup provides carbohydrates for fast population build-up.

Tracheal mites are causing high death rates among colonies. The tracheal mites enter the breathing tubes, or the tracheae, of young bees. Once inside the tracheae, the mites will suck on their blood and block air passage. Its symptoms are very much alike with Nosema. Tracheal mites are more prevalent in winter and in early spring. You may need to contact your county extension agent so they can assess the disease and verify if it is tracheal mites that infected your colonies. Once identified, infested colonies will be treated with Miticur or a special menthol formulation.

Varroa mites suck on the blood of adult bees and enter the bee colony where they will lay their eggs. These eggs, after they have hatched, will suck on the blood of the other bees in the colony. Colonies infested with Varroa mites will die in three to four years. To prevent this from happening, infested colonies should be treated with Apistan. Apistan is a fluvalinate formulation. However, you also need to contact your county extension agent for the identification of the infestation.

Other pests that may hinder productive beekeeping are Wax moths. Wax moths attack beekeeping equipment. The moths lay eggs near the wax combs. When the eggs have hatched, the larvae will begin to burrow through the combs. The combs will then be plastered with feces and webs. Usually, honey bees can protect their colony from moth larvae. If a wax moth attacked a colony and succeeded, this is most probably associated with another problem. The loss of the queen may turn the colony weak that they are not able to fight off the moth larvae. Stored supers are what are usually damaged by wax moths. You can protect stored supers by piling them at a maximum of five hive supers. You will need to cover all cracks and the top of the pile should be applied with paradichlorobenzene crystals. You will also need to put a lid on the pile. The crystals evaporate so you will need to replenish them every once in a while.

Chapter 3: Bee Biology, Equipment, and Bee Installation

Honeybees are the most well known social insect. They live in large, but well-organized, groups of families. As compared with solitary insects, social insects share diverse set of tasks. They are highly-evolved insects that are able to communicate with one another, create complex nests, defend themselves, and divide the work among them. Because of these behaviors, bees are able to successfully exist in colonies.

Any insect species must meet three criteria before it can be called truly social. Number one, it must have cooperative brood care; that is individuals under one roof help each other take care of the offspring.

Number two, a species must have reproductive division of labor. That means most of the individuals in a colony are more or less infertile, and do the colony labor on behalf of the fertile individuals; in our case the Queen. Number three, a species must have overlapping generations; offspring stay in the colony to help parents produce more offspring.

The honeybee has all three of these traits, so it is called *Eusocial* or truly social. All termites and ants are *eusocial,* while only some wasps and bees are. A complex network of behaviors, chemical pheromones, and common nutritional and security needs holds colonies of bees together.

A new package colony begins with about 13,000 bees and grows to about 50,000 if managed properly. In nature, most colonies only reach twenty-five to thirty thousand before swarming and beginning a new colony. If beginning a new apiary for beekeeping, try to build up each colony to a population of around 50,000 to maximize honey production.

In a honeybee colony, there are three types of adult individuals: the worker, the Queen, and the drone. What's noticeable between them is the size difference. By far, most bees in the hive are workers. These infertile females do all the work of the colony. They clean cells, feed larvae, tend to the Queen, build cone, defend the nest, and forage for pollen and nectar.

The other female in the hive is the Queen. She is easy to distinguish from workers because her abdomen is elongated. Usually, each colony has only one Queen. She lays all the eggs in the colony, plus she gives off chemicals called pheromones that regulate the behavior of the bees.

Male bees are the drones. They are as big the Queen, but they have a stockier body shape and cannot sting. They make up only about 10% of the hive population. Drones do no work in the hive. Their only function is to mate with queens from other colonies.

Every member of the colony has specific tasks to do, depending on its age. No one type of bee is more important than the other; each of them is an integral part in making the colony survive and reproduce.

A bee colony's activities depend with the seasons. Bees' activities are affected by the temperature and the season of the flowers. In winter, bees hibernate and you will seldom see them out of their colony. The only time that they will come out is on warm winter days. After the winter season, bees come out at spring time where they can start collecting nectar and pollen from plants. It is also at this time of the year that flowers bloom since their most helpful pollinators, the bees, are out and about.

Equipment for Beekeeping

You will need several pieces of equipment to make your beekeeping more efficient and enjoyable. Obviously, no one likes getting stung, and stinging is something to consider in beekeeping. Fortunately though, bee breeders have worked for decades to produce babies that are gentle and workable. And with enough protective clothing, you can work on your apiary for hours with practically no stings at all.

In the beginning, you may be more confident with a full bee suit with coveralls, extra long gloves, and a veil. However, bee suits are hot. And as you get more confident with bees and the idea of stings, you may be satisfied with just a veil. It's a good idea, though, to keep a full bee suit on hand for those days when the bees are particularly irritable. You will need more protection especially at times when you need to remove hives from their homes. You may need to invest on a hat and a veil, at the very least. A bee sting on your face or your head is not a pretty sight. If you are allergic to them, bee stings are serious. For most beekeepers, a plastic pith helmet that comes along with a veil is a good choice.

Over time, many bee keepers develop a resistance to bee stings, to the point where they don't even swell when they are stung. And they have become experts on removing the stinger without releasing all the venom. You can scrape the stinger off your skin by using your fingernail or a hive tool. A credit card can also do the trick. For most bee keepers, the joy of beekeeping far outweighs the inconvenience of stings.

Your equipment should include a hive tool. Hive tools are specially shaped instruments to pry apart *supers* and frames. Bees glue hive parts together with beeswax and propolis--a mixture of tree resins. The prying end is broad to reduce damage to wooden surfaces. A narrow screwdriver, on the other hand, would cut into the wood.

The other end is a general scraper. The smoker is perhaps the most important beekeeping tool. Smoke has been used for millennia to calm bees and reduce stinging. And after all this time, we still don't know for sure why it works. When bees are smoked, they run deeper into the hive, and many of them start to eat honey out of the cells.

This might be an instinctive response to fire and the bees are preparing to evacuate the nest and start another one. Bees engorged with honey in this way may be less inclined to sting. According to a more sophisticated theory, smoke may mask alarm odors that the bees use to warn each other of an intruder, thereby suppressing the defensive response.

However, these are really academic questions. The bottom line is smokers calm bees and reduce stinging. The earliest smokers were simply smoldering torches. But in 1875, Moses Quinby invented the first practical smoker with a bellows attached to a fire pot.

This is the same basic design we use today. The bellows blows air through the fire pot and directs the smoke out the spout. Dry cow pies, corn cobs, and old burlap make good smoker fuel. However, most bee keepers today like to use pine straw.

Early American beekeeping was never more than a small backyard industry. But with the milestone inventions of the 19th century--the movable frame hive, beeswax cone foundation, a centrifugal extractor, the bellow smoker--large scale beekeeping became possible. With motorized vehicles, bee keepers could move bee hives more easily and maintain several apiaries, instead of just one at home.

Until the 20th century, most honey was eaten at the comb. But in the early 1900s, demand grew for bottled liquid honey that had been extracted from the comb. During the sugar shortages of World War I, consumption of extracted honey skyrocketed. To handle this new demand, bee keepers enlarged their operations. With this growth in beekeeping came a need for reliable source of bees.

Just before World War I, a specialized branch of beekeeping started in the southern states and California--Queen and packed bee production. Today, this multi-million dollar industry mails bees to professional bee keepers and hobbyists all over the world.

Receiving and Installing Packaged Bees

2 to 5 pound packages of bees are routinely shipped by the postal service, but you'll have to pick them up. They don't deliver. If you have to wait before you install your packages,

store them in a cool shaded place. In many parts of the country, many packaged colonies can't produce surplus honey until their second year. However, let's hope that with some good management, you'll get honey from your hives in the same year.

One you receive the package, pry the package crate apart so you can install them individually. Each package holds three pounds, which is over 13,000 bees a piece. Starting this way, instead of already established colonies, allows you to observe the natural build-up of a new hive.

Installing package bees isn't especially difficult. Nevertheless, it can be intimidating to a beginner. First, you need to take out some combs from the *super* to make room for the bees. If you're using *supers* with Langstroth dimensions, take out and temporarily set aside five frames.

Pry off the package lid. Find and remove the Queen in her cage. Examine the Queen as you place her in the hive. Make sure she's not injured and check for general vigor. Suspend the Queen between the two center-most combs.

When you put in the Queen, don't place the screen inward against the frame because the workers on the outside can't reach the Queen and feed her. Instead, place the frame sideways so that the workers have full access to the Queen through the screen.

It's important to spray the packages with sugar syrup, and to spray them real heavily. This makes them sticky, don't fly as much, and easy to pour. After spraying them heavily, shake the bees in the package down into the bottom. Once they're all at the bottom, spray them some more.

After the second spraying, you can now remove the feed can. The bees are now ready to pour into the hive. It is important to shake quite a few bees right on top of the caged Queen. The rest of them can go inside the space you created earlier. Keep bouncing the package to shake out all the bees.

Spread two round tablespoons of Terramycin dust on top of the bees. Terramycin treatments will be repeated two more times at four day intervals. Return combs to the hive as quickly as possible, so the bees will have them to disperse over. This will prevent suffocation.

Move slowly and confidently with your bees. Be sure not to pinch bees between frames. Move them aside gently when necessary. Because of the Queen cage, we won't be able to fit in the 10th frame. Return it later when you remove the Queen cage.

Punch holes in the wax paper wrapping of a pollen substitute patty. This helps the bees chew through the paper and feed. Now place it on top of the frames. Carefully place the inner cover upside down to accommodate the protein patty.

Place grass clippings in the entrance to keep the bees from coming out. Feed your bees medicated syrup continuously until natural nectar flow begin. A variety of feeders are available on catalogs. However, it is preferable to use the Boardman entrance feeder. Repeat the installation process for each of the hives that you have.

Releasing Queens

You'll probably install packaged bees only once a year, making it a unique, yearly experience. You will be entering the hives, however, on a much more regular basis, and you need to become familiar with the techniques for doing that. Always use a smoker when working with bees. Smoke calms them down and makes them a lot easier to work with.

Use an old telephone book or newspaper for kindling, and then add some pine straw and keep puffing so that it catches on fire. After you have the pine straw burning, start adding more pine straw and pack them in tightly. The objective is to create lots of cool, smoldering smoke.

The next thing you need to do is check the Queen and release her. When you're working on a hive, never stand in front of the entrance. You will only irritate your bees if you get in front of their flight path.

First, puff smoke in the entrance. After smoking the entrance, remove the hive covers. Puff smoke across the frames and pull out the Queen's cage. Notice how the bees are attracted to the Queen. If you notice that the other bees are biting the wire of her cage, do not release her yet. Wait another day.

If the bees are trying to feed the Queen with their long tongue through the screen of her cage that is a sign that they're ready to accept her, this is the perfect time to release the queen from her cage. Brush off the bees from her cage and then pry back the wire mesh screen. You have to be careful when you do this because the Queen can fly away.

Slowly peel away the screen and put her in the hive. What you will probably have left hanging on the cage are worker bees. They're still attracted to the cage because it still smells like the Queen. At this point, just shake off the bees and discard the cage.

Now that you have the Queen cage out of there, you now have room to put back in the remaining frames that you took out prior. Puff the bees with smoke just to calm them and then scoot the last frame gently over so you don't have to crush any more bees than you have to. Slide the remaining frame back in so it'll be back to its original ten frame condition.

Put back the inner and outer cover. Your colony now has a free Queen that is ready to start laying eggs. There's another method in releasing the Queen. If this is your first time, we recommend this second method. You see, the Queen is young and nervous. She is also attracted to light. Therefore, she might try to fly away.

So, instead of pulling back the wire screen, you can also remove one of the corks and let the Queen come out on her own. A queen cage has a plug of candy on one end, and nothing on the other. However, there's a cork at both ends. Most people remove a cork on the end with the candy. This lets workers on the outside eat through the candy and then gradually release the Queen.

If you determine that the workers have already accepted the Queen, it is preferable to remove the cork from the end with no candy and release her quickly. After a few minutes, you can come back and remove the empty queen cage. You have to expect to lose a Queen from time to time. You'll have to order a new one right away.

Bee Stings

One thing we haven't talked about yet--and we really need to--is bee stings. If you're going to keep bees, you're going to get stung. And being stung by a bee is really painful. However, you can do a couple of things to minimize the number of stings you get from your bees. For instance, never ever swat at a flying bee. Although bees are gentle creatures by nature, they may sting once they feel threatened or when they sense that their hive is in danger. Bees respond aggressively to fast motions or jarring vibrations.

If you swat a bee, it will only aggravate the entire swarm. Instead, learn to work calmly with your bees. Bees are very busy creatures and they usually will not mind you as they go about their business. Watch experienced beekeepers and see how smoothly they work with their bees; how they brush them aside with their fingers to avoid pinching them for example. Folklore says bees can smell the fear on a novice beekeeper, or that they recognize their owner. But instead, it is really the skill of the beekeeper that cuts down on the amount of stinging. Since there is no way we can dispute these claims, it is important to follow suit with other skilled beekeepers.

If you get stung near a hive, you need to move away very calmly. Do not panic by swatting the bees, waving your arms, or running. If you freak out, you will upset the bees and you will have greater chances of being stung again or being swarmed by other bees. Try to quickly remove the stinger before it releases the pheromones. The pheromones signal the other bees of a nearby threat, which in this case, is you.

When a bee stings you, its poison stack and stinger will remain embedded in your skin. Never pull this poison sack away because that injects all the remaining venom into you. Instead, use your finger nail or your hive tool to scrape it away.

The reality is that the more time you spend time and work with the bees, the more comfortable you will get. However, this does not mean that you will never be stung again. You will only get fewer stings. As time goes by, most beekeepers build tolerance with the pain and the venom that bee stings create. Even though it may still hurt, experienced beekeepers may get less swelling, itchiness, and heat from the bee sting.

Try to see it from the bee's perspective; you need to understand that bees react this way after they let you rip off the roofs of their homes, expose their offspring, and take their nectar, pollen, and honey. For bees, stinging is a suicide mission, so this should not be taken lightly. After a bee sting you, he will leave his stack and stinger and he will eventually die.

Medications

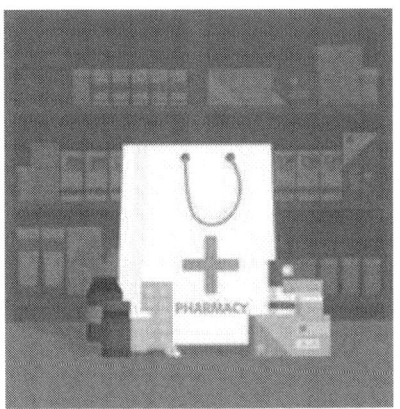

After four days of installing the bees, your colony is now ready for its second Terramycin treatment. Most Terramycin manufacturer recommends three treatments at four-day intervals. It is important to remember that you have to allow at least four weeks between your last treatment and the day you put on your honey *supers*. You don't want to risk contaminating the honey with Terramycin.

So, that's basically how you setup your bee hive. It doesn't need to be difficult. For the first time, the hardest part is opening up a package of live bees. However, after you cross that hurdle, your apprehensions will grow less and less as you learn to appreciate the wonders of a bee colony.

All we did in this chapter was to provide the bees and nest site, and temporary food. The bees still have to accept their queen, build comb, begin rearing brood, scout out local nectar and pollen sources and ultimately, produce surplus honey for us to harvest. It's your job to help them along.

Chapter 4: Bee Management

The main goal of bee management is to maximize colony populations before the major nectar flows begin. If their population needs are satisfied, bees turn the nectar into surplus honey that we can harvest, instead of using it to make more bees. Depending on which part of the country you live in, nectar may come either during the months of April or May, which doesn't leave you much time.

You must push the colonies to grow fast, or you may miss an entire year's worth of honey. In the last chapter, we installed the packed bees in their hives. And these young colonies are now bursting with activity. Workers are building combs and the Queens are laying eggs non-stop.

You must feed the colonies, medicate them, and watch closely for any problems. In this chapter, we'll take a look at the normal development of colonies and how to manage them. During this time, we'll see the first few workers emerge--we'll see the population swell. And before we know it, it's time for the nectar flow.

We'll add honey *supers* and even move the hive to another area to collect a very special kind of honey: sourwood. Managing a colony is a very busy time for a beekeeper, so sit back , enjoy and let's see what happens.

Hive Maturation

Before you even enter the hive, look at the entrance. Bees are bringing in large amounts of pollen. Notice the balls of pollen on their legs. The hind legs of a worker have a structure called the pollen basket.

When a bee visits a flower, it collects pollen with its front legs and mouthparts. Pollen is passed down to the rear legs and forced into the pollen basket. Back at the hive, the bee removes the pollen pellets using its middle legs, places it in a cell, and the house bees pack it in tightly.

It is always a good sign when you see bees collecting pollen. This usually means the Queen is well and the colony is rearing brood. Queen-less bees without brood are not motivated to collect pollen.

Before we go any further, let's talk about the development stages a young bee goes through. First, the Queen lays an egg in the bottom of the cell. Three days later, it will hatch into a tiny larva. Workers feed the young larvae with pollen and a secretion called brood food. They grow quickly, and when they fill most of the cell, the workers cap the cell.

The larvae then transforms into a pupa. This is a quite non-feeding stage that somewhat resembles the future adult. Several days later, the pupa molts into an adult and emerges from the cell. Development times for workers, drones, and Queens vary.

One week after your release the Queen into your colony, check the colony growth. By this time, there should be white wax combs under construction. These cells contain sugar syrup, eggs, and larvae. Make sure to check the Queen. Notice how she stands out from the workers because she is longer, her thorax is bigger, and her abdomen is elongated and looks like leather.

Mature Queens move slowly around the combs as they look for empty cells in which to lay eggs. At first, you may have trouble finding the Queen. But with experience, you'll find her more quickly. When you work a hive, assume the Queen is on every frame you pick up. That way, you will work more gently and reduce the chance of accidentally killing her.

While you're working the hive, you need to make sure there's plenty of protein supplements and sugar syrup. As your colony grows, you have to add a second hive body, and eventually some honey *supers*.

You'll know that a colony is not doing well when the bees are loud, they act as if they're nervous, and there are no eggs or larvae in the hive. This might also mean that the Queen is dead. You still have time to salvage a colony with a dead Queen, but you're going to have to move quickly and get another Queen in there. You can order another Queen from your trusted bee supplier.

While your order is still on its way, you can take a frame from a strong colony, shake off all the worker bees, and slide it in the Queen-less colony. When taking a frame, make sure that you don't accidentally take the Queen of that colony.

Stocking brood in this Queen-less colony will help boost its population and maintain it until the new Queen arrives. It is never a good idea to steal brood from a colony, but the benefits to the Queen-less colony far outweigh the costs to the donor colony.

The Brood Nest

Two weeks after you release the Queen into the colony, check brood development again. On the first week we found that new wax cells have been built and were filled with eggs, young larvae, and sugar syrup. On this second week, you may find many more eggs, larvae, and something else: capped brood.

As the larvae mature, the installed bees cap the larvae cells with a wax seal. After around two weeks, the larvae transform into adults, similar to caterpillar in a cocoon. Young bees will emerge from these cells in about another week.

The sealed brood looks like brown cardboard. The sealed syrup is easily distinguished from the brood. Eventually, the syrup will get used and will be replaced with nectar, which will become honey.

A good Queen in a colony can be characterized by her meticulous pattern of filling every cell with an egg, rarely skipping one. A Queen that is not as meticulous can be characterized by the irregular brood pattern it creates. We build our *supers* following a square box design for our convenience. However, the bees build their nest in a spherical design.

At the center is sealed brood. It's sealed because this is where the Queen began laying eggs first. She then radiates outward as she continues to lay. Therefore, the brood is progressively younger as you move outward. The cycle will begin again as the brood in the middle begins to emerge. The Queen will then begin laying eggs in the center again.

Above and to the sides of the brood is a thin layer pollen. This protein fuels brood production. The larvae are fed this pollen. Nurse bees also eat the pollen, which stimulates them to secrete the brood food that they feed each larvae.

The bees will fill these frames with either honey or sugar syrup. Some of the cells will be capped, and some will not. This is because bees evaporate excess moisture from the liquid. This is done by fanning the cells with their wings. When the moisture content is reduce to 18%, they will cap the cells. These serve as the carbohydrate, or energy stores, for the colony.

Three weeks after you release the Queen into the colony, go back to the hive and observe their development. Since it takes 21 days for bees to develop from an egg to a young adult, you may have some bees emerging from their cells. That is assuming that the Queen started laying eggs on her first day.

Remember that during the producing season, the life span of these honey bees is only six weeks. It may be as little as three weeks. An emerging bee chews through her capping and climbs out without any help from other bees. They are light-colored and their hairs are matted together.

For a few hours, her exoskeleton, or cuticle, is very soft. So soft, in fact, that she cannot sting. As a bee ages, it performs a series of tasks in a predictable sequence. Here they are in sequential order:

1. Capping cells of brood
2. Clean cells
3. Tend to the brood
4. Tend to the Queen
5. Receive incoming nectar
6. Clean out hive debris
7. Take their first orientation flight
8. Build comb
9. Receive incoming pollen

10. Ventilate the hive with their wings

11. Guard the entrance

12. Begin foraging for nectar and pollen

Foraging is the most dangerous job because bees expose themselves to predators outside and bad weather. At this stage, bees literally work themselves to death, and you will find the ground in front of your hives littered with their dead and worn out bodies.

We have intentionally waited three weeks because bees prefer to build vertically, rather than horizontally. They would have built up only the center comb and moved up into the second hive body, before they would have built out all the columns in the first hive body. We want them to occupy the first hive body, because that encourages construction of a large brood nest and consequently large populations.

Seven weeks after your colonies have been in place--that's four weeks since you added the second hive body--notice how the bees have maintained the spherical shape of their nest even into the second hive body. Not all beekeepers use two hive bodies. In the southern parts of the United States, beekeepers just use one because bees don't need as much space for a large winter food supply.

With just one hive body, hives are lighter and easier to move. It is also easier to find the Queen and medicate. On the other hand, two hive bodies supply abundant space for rearing brood and storing food. It really boils down to your personal preference.

An intensive feeding and medicating program will promote fast population growth. The bees will collect enough nectar for themselves and will be strong enough to start collecting surplus nectar for us. The incoming nectar that the bees will store in the shallow *super* is our harvestable honey. We won't take honey from the hive bodies because bees will need it later in the year.

The Queen might move in to the honey *supers* and lay eggs alongside cells of nectar. That's not really a problem because you can filter out the brood during the harvesting process. However, many beekeepers don't like brood in their honey *supers*. If you plan on using a queen excluder, install it between the hive body and the honey *supers*.

As the name implies, the queen excluder restricts the motion of the Queen. Workers can pass through the queen excluder's grates. However, the drones and Queen are not going to be able to do so because they are too large. Some beekeepers don't use queen excluders because it slows down the process of storing honey.

If you're going to use frames of drawn foundation (the foundation built by the bees themselves) allow extra space between the frames. The bees use this extra space to build thicker combs, which makes it easier to harvest honey.

It is also important to make sure that the honey would not be contaminated by the medications you were using. Since we're at week 7, our Terramycin dusting was more than five weeks ago. And if your remember, the manufacturer recommends allowing at least four weeks between the last application and putting on the honey *supers*. At this point, we know that we are safe.

We've also been feeding the colony sugar syrup with Fumidil-B. The manufacturer of that medication says don't use it immediately before or during a honey flow. To be extra safe, we stopped over two weeks ago when our syrup supply ran out. So, we can be rest assured that our honey will be pure, wholesome, and free of any contamination from the medications.

Earlier, we were talking about the incoming nectar. Nectar is the sweet liquid secreted by flowers of various plants. Bees gather the nectar and turn it into honey. Nectar determines the flavor of the honey. Let's say for example that in a particular area, there are a lot of clovers. If the bees were to gather the nectar of these clovers, we would have clover honey.

Knowing when there's nectar close is almost second nature to a veteran beekeeper. These are the signs which indicate there is a nectar flow:

1. The bees aren't drinking syrup because they are getting nectar.
2. They are very gentle and very industrious because they are pre-occupied with their work.
3. There are many bees festooning--they hook together and form long chains.
4. Wax scales can be found on honey bee bellies.

5. White wax can be found on frames.

6. Bees exhibit no robbing behavior because they have enough of their own.

Chapter 5: Harvest Time

In this chapter, we'll see how to harvest and process honey; Nature's most delicious sweet. When it comes to flavor, freshness and purity, we cannot improve upon honey in its natural state in the comb. But nowadays, most people prefer liquid honey, so a little processing is necessary. Remember, however, that as we process honey, we must conserve as much as possible the perfect condition in which bees left it.

Harvesting and Processing

Now we begin the process of harvesting honey--the golden nectar. We'll be removing honey *supers* that are full and capped, that is those at the bottom of the stack that we put on first. When you first crack the lid of the hive, the smell of honey fills the air. With experience, you'll be able to distinguish the type of honey by its distinctive aroma.

The honey cells are capped with new white wax, which the worker secrete. Take only the honey that is mostly capped. Otherwise, you risk the chance of having your honey ferment because of excessive water content. You see, after the bees have deposited the nectar in the cells, they fan air over it to evaporate out the excess moisture.

Only when the nectar is at the 18% moisture level of ripe honey, will the bees cap the cells. To harvest honey, first we must remove the bees from the *supers*. If possible, try to use bee repellant and a fume board. Sprinkle some repellant on the fume board. Then, remove the hive lid and put the fume board in its place. The repellant drives the bees down into the lower *supers*.

After about five minutes, the *super* is cleared of most bees and can be removed. When bees smell honey, it may start a feeding frenzy, which may even make them start robbing neighboring colonies. You need to cover the *supers* to reduce the chance that they will go for the exposed honey.

Keep the honey *supers* in a clean, warm, and dry location until you can process the honey. Because it will absorb moisture from the air, it is important to extract the honey as soon as possible to maintain the quality of your harvest.

Moisture content is one of the most important factors influencing the quality of your honey. Honey with high moisture is thin and unappealing, and has a high risk of fermentation. To thicken it, you'll have to dehydrate the honey while they are still in the comb. That's the easiest way to do it. And even though most of the honey was capped, its quality can still be improved by dehydration.

As much as possible, try to move the honey *supers* into a clean, dry utility room that can be closed off. Stack the *supers* on empty hive bodies in a staggered fashion. This is done to allow airflow between each frame. Use a dehumidifier to draw out moisture, and a fan to circulate the warm, dry air over the honey. If you keep the combs warm before extracting, the honey flows easier and harvesting is much more efficient.

The longest you want to dehydrate the honey is one week. The shortest is two days. You'll have to use your better judgment to determine how long to dehydrate your honey. However, this only comes with experience.

Extracting Honey

After drying your honey for several days, you can be confident that you can extract it and get the quality that you want. There's a good reason for drying your honey: uncapped cells. Almost every frame will have some uncapped cells. In fact, in any *super* there might be as much as one-third of the cells uncapped.

If you remember back when we put honey *supers* on the hives, we remove the 10th frame from *supers* with drawn out foundation. Using only nine frames allowed the bees to draw out thicker combs that are much easier to uncap. In fact, some people even think that you get more honey from nine thick frames of comb than from ten frames.

We need to follow a very specific sequence to extract honey. First, put a frame on a nail point to allow it to pivot. And then, using a heated capping knife, slice open the cells of honey on each comb. A word of caution: The dripping wax is scalding hot. The capping will fall into a screening basket in an uncapping tank. The honey will continue to drip out of the capping into a catch bucket.

Place the frames of uncapped combs on the tank next to the honey extractor until there enough to load. This will keep the honey dripping into the tank. A centrifugal force extractor throws honey out of the cells onto the tank wall. A non-reversible honey extractor can remove honey from only one side of the comb at a time.

More advanced radial extractors can remove honey from both sides of the comb simultaneously. Sometimes, combs will break while being spun, but there is a technique to spinning that will save many of the combs. After spinning the first side gently for about 30 seconds, or until about half of the honey has been spun out of the cells, reverse the frames.

Now, extract the second side completely. Then, turn the frames back to the first side and spin out the remainder of the honey. When all the honey is out, put the frames back in the *supers* and use the same combs next year. The honey in the extractor contains dead bees and bits of wax, most of which is removed as it passes through the double screens.

Two *supers* will usually have enough honey to fill a five gallon bucket. From the extractor, the honey is poured into a settling tank where it will sit undisturbed for at least a week. This lets small particles of wax and air bubbles rise to the surface where you can skim them off. Air bubbles can cloud the honey and cause a scum on top of the jar.

Most commercial honey producers would heat their honey to aid in the final straining. This is done to prolong shelf life by killing yeast that promote granulation. However, heating is not really necessary for most hobbyists. All honey will granulate, but can always be re-liquefied by placing the jar in boiling water.

The harvesting process yields not only honey, but also plenty of beeswax. The shaved off cappings that fell into the uncapping tank are new wax that was produced by the bees. And because this wax contains few contaminants to darken the final product, it is the highest quality and the most desirable of all beeswax.

Conclusion

By now you may feel overwhelmed with all the intricacies of keeping honey bees. However, you don't need to know before you begin. In fact, you can't know everything before you begin. All of us learn as we go, and the best beekeepers are those that keep on learning new things.

You can also join beekeepers club so you can learn more from the experience of other beekeepers and they will learn from you as well. You should also keep your expectations low on the first year as you are just learning the ropes. Do not get discouraged by the mistakes that you will make; you will learn more from them. You should not expect large profit on your first year, although you can hope for it. After you get past the first year of your beekeeping, it should be so much easier for you then. As you go along, you will discover what you need to improve for a better harvest the following year. The learning you will gain from your experience far outweigh any manual you may have read. Experience will be your best teacher. And besides, nothing is more rewarding than seeing the fruits of your labor come to life and multiply.

Thank you for purchasing this book and we hope that we've taught you a lot on how to start your very own beekeeping operation.

Also if you enjoyed this book, then I'd like to ask you for a favor, would you be kind enough to leave a review for this book on Amazon? It'd be greatly appreciated!

Made in the USA
Middletown, DE
31 May 2017